George Windle Read

**The Automatic Instructor**

A practical system for home study

George Windle Read

**The Automatic Instructor**
*A practical system for home study*

ISBN/EAN: 9783337036249

Printed in Europe, USA, Canada, Australia, Japan

Cover: Foto ©berggeist007 / pixelio.de

More available books at **www.hansebooks.com**

# THE

# AUTOMATIC INSTRUCTOR,

### A

## PRACTICAL SYSTEM FOR HOME STUDY.

ST. PAUL, MI.
WM. KENNEDY PRINTING CO.
1898.

# CONTENTS.

# THE
# AUTOMATIC INSTRUCTOR.

## INTRODUCTION.

Some years ago I was very anxious to obtain a certain appointment to be made by competitive examination. One of my friends whom I consulted about the matter and who had previously won a similar appointment, told me that the key to success in a competitive examination was a determination to win, and that with such a determination, backed by persevering and diligent study, failure would be unlikely if not impossible.

It occurred to me that if the other candidates were also imbued with this spirit of determination, success would fall to the one who was most earnest and careful in his preparation. I made up my mind no one should surpass me in this respect.

After studying hard for a week or more, I became discouraged. My progress in no sense corresponded with the efforts I was making. This, I now see clearly, was due to the following circumstances :

1.   I had been out of school for a long time and my mind had lost the training acquired by the habit of daily study.

2.   In the absence of an instructor to guide them, my efforts were not sufficiently systematic and the actual progress made was uncertain.

3.   In my anxiety for thorough prepara- tion in a limited time there was a constant tendency to hurry and an effort to learn more in one day than the mind could possi- bly assimilate.

This led to mental fatigue and mind wandering. I often became conscious of having read a dozen pages or more while thinking of what I would do after getting the appointment or of something else entirely foreign to the contents of the book.

It also led to the belief that my memory was weak, for a carefully studied lesson would soon become confused and indistinct. The conviction that I was wasting time of which every moment was precious led to a careful consideration of the case and to the discovery of a system so simple, so easy of application, and so effective, that I was able to learn with absolute thoroughness the various subjects to be covered by the examination.

I attribute my success in the competitive examination, and in obtaining the coveted appointment, entirely to the system described in the following pages. While especially adapted to use in preparing for examinations, it is believed this system will be found of value by anyone pursuing a course of study or reading at home, without an instructor.

G. W. R.

# I

# GENERAL PRINCIPLES OF THE SYSTEM.

———

Every practical system for learning a book is based on the following fundamental principles :

*a.* The mind must be concentrated upon what is read.

*b.* An analysis must be made, separating the essential from the immaterial.

*c.* The impression produced by the essential must be strong enough to cause the mind to retain it, or the original impression must be sufficiently strengthened.

These three principles may be summarized in the three words.

*a.* Concentration.

*b.* Analysis.

*c.* Retention.

It is well known that the most natural way to deepen a mental impression is by *repetition.*

If part of a lesson is perfectly known, repetition should be confined to the part remaining to be learned, so as to impress the mind most strongly with that which is assimilated most slowly. This makes it desirable to *separate* the parts which the mind grasps strongly at once and retains firmly from those grasped less strongly and retained with difficulty.

After the learning process is finished, its thoroughness should be tested by an *examination;* otherwise one cannot be sure of having accomplished his purpose.

In the system to be explained, *retention* is assured by :

*d.* Repetition.

*e.* Separation.

*f.* Examination.

and the elements of the system may be completely classified as follows :

*a.*   Concentration.

*b.*   Analysis.

*c.*   Retention.   $\left\{\begin{array}{l} d. \\ e. \\ f. \end{array}\right.$   Repetition.
Separation.
Examination.

Concentration and analysis go hand in hand.

Every book to be learned is made up of essential points usually preceded or followed, or both, by explanations or illustrations and other incidental matter intended to elucidate the main principle or point.   The salient points form the "meat" or substance of the book, and when the mind has grasped, digested, and retained them, the book is learned.   After being read and understood, it is neither neces- sary nor desirable to burden the memory with the subordinate matter, which is only useful in helping the reader to understand the main points and in impressing them upon his mind.

Books differ greatly in regard to the relative number of these material points. In a text-book, for example, nearly every sentence will contain some statement or principle necessary to be learned and remembered. In popular scientific and historical works, the illustrative and explanatory matter will be proportionally greater, while in works of fiction the reader may find few, if any, important points which he will care to learn and remember.

In a text book, little is left to choice in the way of selecting the essential points. They follow one another like the steps of a stairway leading from ignorance to knowledge of the subject treated.

In the reading of popular educational works, and works of fiction, the choice of the points to be remembered may be affected by the taste or special object of the reader.

But whatever the book read or studied, the process of learning it will consist in an analysis of the contents and the retention

of this analysis in the mind of the reader
or student.   Without concentration, both
analysis and retention are impossible ;
without analysis, retention is impossible ;
without retention, reading or study is labor
lost.    Hence, all these are necessary to the
mastering of any book.

# II.

# GENERAL DESCRIPTION OF THE SYSTEM.

The simplicity of the system referred to in the Introduction enables it to be described in a few words :

The first step consists in formulating questions covering every point of im-portance in the text and in writing these questions on cards or slips of paper, with the number of the card and a reference to the place where the answer is to be found.

The second step consists in taking the cards containing the questions prepared during the first half of the daily period available for study, and in devoting the

second half of this period to answering
these questions without reference to the
book. Whenever the proper answer is in
doubt, the card containing the question is
laid aside until the conclusion of the first
attempt to answer all the questions. Then
the answers to the questions on all cards
which have been so laid aside are looked
up, after which a second attempt to answer
them is made without reference to the
book, laying aside as before, the card containing any question still answered doubt-
fully. This process is continued until all
the questions prepared can be promptly
and correctly answered without reference
to the book.

The third and final step consists first, in
applying the process described in the pre-
ceding paragraph to all the cards prepared on the book and second, in applying
the same process after thoroughly shuffling
the cards.

## I I I.

## THE SYSTEM APPLIED.

———

When the essential points are perfectly clear, as is usually the case in a text-book, the application of the system is very simple. For example, suppose it is desired to acquire a thorough knowledge of the "Constitution of the United States." Take the book containing the Constitution, a pencil, and a number of cards or slips of paper of uniform size, about an inch and a half wide by four inches long.

The Constitution consists of a Preamble followed by a number of Articles divided into Sections, which, in turn, are divided into Clauses. Upon reading the Preamble, it is seen that the Constitution was adopted by a specified people for certain defined objects. It is necessary to remember by whom it was adopted and for what purpose ; therefore take one of the cards, number it in the upper left hand corner, and write a question, the answer to which is the essential point to be learned. Then place a reference in a convenient place on the card showing where the answer is to be found. Proceed in like manner with the succeeding Articles, sections, and clauses, as follows :

## CONSTITUTION OF THE UNITED STATES.

We, the People of the United States, in order to form a more perfect union, establish justice, insure domestic tranquillity, provide for the common defense, promote the general welfare, and secure the blessings of liberty to ourselves and our posterity, do ordain and establish this Constitution for the United States of America.

(1)

Constitution U. S. Preamble.

———

What were the six objects of the Constitution ? By whom was it ordained and adopted ?

ARTICLE I. LEGISLATIVE DEPARTMENT.

SECTION I. All legislative powers herein granted shall be vested in a Congress of the United States, which shall consist of a Senate and House of Representatives.

(2)

Constitution U. S. Article I. § I.

———

What body is vested with the legislative power granted by the Constitution ? Of what does this body consist ?

SECTION II.—*Clause 1.* The House of Representatives shall be composed of members chosen every second year by the people of the several States, and the electors in each State shall have the qualifications requisite for electors of the most numerous branch of the State Legislature.

---

(3)

Constitution U. S. Article I. § II.
Clause 1.

—

What is the composition of the House of Representatives ?

What qualifications must be possessed by the electors in each State ?

---

*Clause 2.* No person shall be a representative who shall not have attained to the age of twenty-five years, and been seven years a citizen of the United States, and who shall not, when elected, be an inhabitant of that State in which he shall be chosen.

(4)

Constitution U. S. Article I. § II.

Clause 2.

———

What three qualifications must be possessed by a representative ?

*Clause 3.* Representatives and direct taxes shall be apportioned among the several States which may be included within this Union, according to their respective numbers, which shall be determined by adding to the whole number of free persons, including those bound to service for a term of years, and excluding Indians not taxed, three-fifths of all other persons. The actual enumeration shall be made within three years after the first meeting of the Congress of the United States, and within every subsequent term of ten years, in such manner as they shall by law direct. The number of representatives shall not exceed one for every thirty thousand, but each State shall have at least one representative ; and until such enumeration shall be made, the State of New Hampshire shall be entitled to choose three ; Massachusetts, eight ; Rhode Island and

Providence Plantations, one ; Connecticut, five ;
New York, six ; New Jersey, four ; Pennsylvania,
eight ; Delaware, one ; Maryland, six ; Virginia,
ten ; North Carolina, five ; South Carolina, five ;
and Georgia, three.

---

(5)

Constitution U. S. Article I. § II.
Clause 3.

———

How are representatives apportioned among the several
States ?

What else is apportioned on
the same basis ?

---

(6)

Constitution U. S. Article I. § II.
Clause 3.

———

In the apportionment of representatives how are the "respective
numbers" in a State determined?

What actual enumerations are
provided for ?

(7)

Constitution U. S. Article I. § II.
Clause 3.

———

What are the limits of the number of representatives from each State ?

How many members had the first House of Representatives ?

*Clause 4.* When vacancies happen in the representation from any State the executive authority thereof shall issue writs of election to fill such vacancies.

(8)

Constitution U. S. Article I. § II.
Clause 4.

———

When vacancies happen in the representation from any State, how are they filled ?

*Clause 5.* The House of Representatives shall choose their Speaker and other officers ; and shall have the sole power of impeachment.

---

(9)
Constitution U. S. Article I. § II. Clause 5.

What is the name of the presiding officer of the House of Representatives and how are all officers of that body chosen ? What body has the sole power of impeachment ?

---

SECTION III.—*Clause 1.* The Senate of the United States shall be composed of two Senators from each State, chosen by the legislature thereof, for six years ; and each senator shall have one vote.

(10)   Constitution U. S. Art. I.
§ III.  Clause 1.

——

What is the composition of the
U. S. Senate ?
How are senators chosen ?
For how long ?
How many votes has a Senator ?

*Clause 2.* Immediately after they shall be
assembled in consequence of the first election,
they shall be divided as equally as may be into
three classes,  The seats of the senators of the
first class shall be vacated at the expiration of
the second year ; of the second class at the expiration of the fourth year ; and of the third
class, at the expiration of the sixth year ; so that
one third may be chosen every second year ; and
if vacancies happen by resignation, or otherwise,
during the recess of the Legislature of any State,
the executive thereof may make temporary ap-
pointments until the next meeting of the Legislature, which shall then fill such vacancies.

(11) Constitution U. S. Art. I.
§ III. Clause 2.

---

Upon assembling after the first election, how were the senators to be classified ?

What was the object of this classification ?

(12)

Constitution U. S. Article I.
§ III. Clause 2,

---

When a vacancy in the Senate occurs during a recess of the Legislature of any State, how is it filled ?

*Clause 3.* No person shall be a senator who shall not have attained to the age of thirty years, and been nine years a citizen of the United States, and who shall not, when elected, be an inhabitant of that State for which he shall be chosen.

(13)

Constitution U. S. Article I.
§ III. Clause 3.

---

What three qualifications must a senator possess ?

*Clause 4.* The Vice-President of the United States shall be president of the Senate, but shall have no vote, unless they be equally divided.

*Clause 5.* The Senate shall choose their other officers, and also a president *pro tempore*, in the absence of the Vice-President, or when he shall exercise the office of President of the United States.

(14)

Constitution U. S. Article I. § III.
Clause 4 and 5.

Who is president of the Senate?
How are the other officers
chosen ?
When and how is a president
*pro tempore* chosen ?

*Clause 6.*  The Senate shall have the sole power
to try all impeachments ; when sitting for that
purpose they shall be on oath or affirmation.
When the President of the United States is tried,
the Chief Justice shall preside ; and no person
shall be convicted without the concurrence of
two thirds of the members present.

(15)

Constitution U. S. Article I. § III.
Clause 6.

———   .

Where does the sole power of
trying impeachment rest ?   Who
presides during a trial of the
President of the U. S. ?   What
is essential to conviction ?

*Clause 7.*  Judgment in cases of impeachment
shall not extend further than to removal from
office, and disqualification to hold and enjoy any
office of honor, trust, or profit, under the United
States ; but the party convicted shall neverthe-
less be liable and subject to indictment, trial,
judgment, and punishment, according to law.

(16)

Constitution U. S. Article I.
§ III.    Clause 7.

How far may judgment extend in cases of impeachment?

Does this bar further proceedings and subsequent trial and punishment?

It will be assumed that only one hour was available for study at this time and that half an hour has been consumed in formulating these questions and preparing the sixteen cards.

Closing the book, the cards, arranged in a pack in numerical order, are taken in hand, and an effort made to answer the questions.    Whenever the questions on the top card can be answered without difficulty, that card is shifted to the bottom of the pack; but whenever a question is met which cannot be answered without hesitation, the card containing it is laid aside.

It should not take more than ten minutes to run through the cards in the manner described.  At the end of that time, it may be found that half or even more of the cards have been laid aside because of an uncertainty as to the proper answer to some question, or entire inability to answer it.

These cards are now taken and the correct answers looked up in the place indicated by the reference.  The book is then closed again, and another effort made to give the correct answers to these questions.  Should any answer be still doubtful, the card containing the question is laid aside as before, and this process continued until all the questions have been correctly answered.

If any time remains, all the cards may be gone over again, which should not take more than five minutes.

Finally, the cards are arranged in numerical order and fastened together by a rubber band.

The next time study is resumed, proceed in a similar manner and continue from
day to day until the entire Constitution has
been studied.

The cards prepared will now number
about a hundred. The next step is to take
all these cards, arranged in a pack in numerical order, and go over them as described in the case of the first sixteen, being
careful to lay aside every card containing
a question not answered with readiness and
ease.

The answers to these questions are then
to be looked up as before described, and
the process continued until every question
has been answered without hesitation.

Finally, the cards are to be thoroughly
shuffled, so as to arrange them in haphazard order, and are then to be gone
over as before.

At the conclusion of this process, the
student, no matter how mediocre his
natural ability, will be able to pass a
perfect examination on the Constitution.

Moreover the learning process will have been at all times easy and pleasant. The time consumed will not have been great in proportion to the results obtained. If the same number of hours had been devoted to reading the Constitution over and over again, no matter how carefully, the knowledge acquired would have been incomparably less complete than that resulting from the application of this system.

If it is desired to review the subject at any future time, it is only necessary to take the cards originally prepared and go over them as indicated  The entire subject can thus be thoroughly reviewed in a fraction of the time required by any other method.

It will be observed that in preparing the questions the mind is necessarily *concentrated* on the material points.  The process is mechanical.  No effort whatever is required to prevent *mind wandering*.

The preparation of the questions also requires a mental *analysis* of the subject.

Take clause 3 of Section II, for example. A careful reading shows the essential points to be :

1.   The manner of apportioning repre-sentatives in Congress among the several States.

2.   The manner of apportioning direct taxes among the several States.

3.   The manner in which the respective numbers in the several States shall be de-termined as a basis for representation and direct taxes.

4.   The actual enumerations to be made and the manner of making them.

5.   The maximum and minimum num-ber of representatives from any State.

6.   The designation of the numbers of representatives to which each of the orig-inal States was to be entitled prior to the taking of the first census.

The last point might be covered by

questions to bring out the number of representatives allowed each State, but it would be useless to burden the memory with such details. It should be quite sufficient to know the total membership of the first House of Representatives.

In a similiar way a mental analysis is made every time a question is prepared, and the essential point is emphasized by writing the question on the card.

This *concentration* and *analysis*, in themselves, strongly impress the essential points on the mind.

Answering the questions on the cards is the *repetition* necessary to *retention*.

Laying aside the cards containing questions which cannot be answered is the *separation* of the unknown from the known.

Looking up the correct answers to only those questions which could not be answered, impresses the mind strongly with whatever remains to be learned.

Finally, going over all the cards, arranged first in numerical and then in haphazard order, is both a general review of the subject and the *examination* necessary to assure the student of his thorough knowledge thereof.

As another illustration of the use of the System let us take " Algebra," a representative text-book of the conventional type. ____

## CHAPTER I.*

1. *Quantity and Number.* Whatever may be regarded as being made up of parts like the whole is called a *quantity.*

In other words whatever admits of division into parts all the same in kind as the whole, is a quantity.

To measure a quantity of any kind is to find how many times it contains another known quantity of the same kind.

A known quantity which is adopted as a standard for measuring quantities of the same kind is called a *unit.*

Thus the foot, the pound, the dollar, the day, are units for measuring distance, weight, money, time.

A *number* arises from the repetitions of the unit of measure, and show how many times the unit is contained in the quantity measured.

* From Wentworth's College Algebra.

Concentrating the mind upon this para-graph and analyzing it, the essential points may be covered by the following questions.

(1)

### P. 1.   Par. 1.

---

What is quantity ?

How is quantity measured ?

What is a unit ?

What is a number ?

Proceeding in like manner, we have

PAR. 2. When a quantity is measured, the result obtained is expressed by prefixing to the name of the unit the number which shows how many times the unit is contained in the quantity measured.

The result is called the measure of the quantity. The number which shows how many times the unit is taken is called the numerical measure of the quantity.

Thus 7 feet, 8 pounds, are respectively measures of distance and weight; the numerical measures being respectively 7 and 8.

(2)

P. 2   Par. 2.

———

How is the result obtained by measuring a quantity expressed ?
What is this result called?
What is the numerical measure of a quantity ?

PAR. 3.  For convenience, numbers are represented by symbols.  In arithmetic the symbols 0, 1, 2, 3, 4, 5, 6, 7, 8, 9, and combinations of these symbols, are embloyed to represent numbers. The series 0, 1, 2, 3, ......, obtained by counting is called the natural series of numbers.

Any figure or combination of figures represents one, and but one, particular number.

(3) .

P. 2. Par. 3.

—

How are numbers represented for convenience ?

What is the natural series of numbers ?

PAR. 4. *Numbers in General.* Numbers possess many general properties, which are true, not only of a particular number but of all numbers.

For example, the sum of 12 and 8 is 20, and the difference between 12 and 8 is 4. Their sum added to their difference is 24, which is twice the greater number. Their difference taken from their sum is 16, which is twice the smaller number.

We shall see later on that these are general properties of numbers, viz :

The sum of two numbers added to their difference is twice the greater number; the difference of two numbers taken from their sum is twice the smaller number. Or,

(1) (greater number + smaller number) + (greater number − smaller number) = twice greater number.

(2) (greater number + smaller number) — (greater number — smaller number) = twice smaller number.

But these statements may be very much shortened; for, as greater number and smaller number may mean any two numbers, two letters, as $a$ and $b$, may be used to represent them ; then 2a will represent twice the greater number, and 2b twice the smaller. Then these statements become :

$$(1) \quad (a + b) + (a - b) = 2a.$$

$$(2) \quad (a + b) - (a - b) = 2b.$$

In studying the general properties of numbers, letters used to represent numbers may represent any numerical values consistent with the condition of the problem.

(4)

P. 2.   Par. 4.

—

What is meant by "general properties" of numbers?

(5)

P. 2. par. 4.

———

Give example of general properties of numbers based upon relation between sum and difference of two numbers, and express the statement of the general property in three ways ?

(6)

P. 2. Par. 4.

———

In studing the general proper-ties of numbers, what numerical values may be represented by the letters used to represent numbers?

# IV.

## FORMULATING QUESTIONS.

The value to the student of a series of questions carefully prepared to cover the essential points of a subject is fully recognized by authors of text books. A list of such questions is not infrequently found at the bottom of each page, or at the end of each chapter, or sometimes at the end of the book. As a rule, the questions are intended more for the use of the student than for the teacher, and the object is evidently to emphasize the points essential to a thorough knowledge of the book.

In the class-room, questions are used to bring out the important points of the lesson quite as much as to test the student's knowledge. A lecture is often closed by a "quiz." In examinations of all kinds the student's knowledge is more often tested by questions than in any other way.

These questions are prepared by author or instructor. Perhaps the most important step in the system herein described is the preparation of an exhaustive series of questions *by the student*. In the framing of these questions, the important points of the book are introduced to the mind of the student individually and the acquaintance is cemented and catalogued by the process of writing the question and the reference on the card.

Should the system be applied to a book containing questions already prepared, those of the author should not be used until after the student's own questions have been formulated. The latter may then be compared with the former, and modified, if by so doing they will be improved.

In certain text-books, especially those

pertaining to mathematics, pure or applied, the demonstration of theorems and the solution of problems are most important. The solution of problems is merely an adjunct to the system herein described, but should never be neglected by the student. After he has solved all the problems, one or more of them should be selected and copied on a card as in the case of the questions, to be solved during the reviews and examinations prescribed.

A theorem to be demonstrated presents no difficulty. Take for example the following :

## PROPOSITION XVII. THEOREM.*

In any triangle, the straight line which bisects the angle at the vertex, divides the base into two segments proportional to the adjacent sides.

---

* Davies' Legendre.

(105)

Proposition XVII.  Book IV.

---

How is the base of any triangle divided by the straight line which bisects the angle at the vertex ? Demonstrate.

Here the answer to the question is the theorem to be demonstrated, and a knowledge of the subject requires the ability to make the demonstration.

Whenever a rule is covered by a question, the question should be followed by " Why ? ", and the student should invariably be able to give the reason.

In the study of languages, this system is chicfly valuable in learning the principles and rules ; the exercises of course must be worked out separately ; but, as in the case of the problems, it is well to make out occasional cards containing parts of exercises suitable as illustrations.

# V.

# MEMORY TRAINING.

It is claimed that this system trains the memory by the most natural and therefore the most logical of processes. Weak memory results from the inaptitude of the mind to receive strong impressions in ordinary cases.

The remedy is evidently either the removal of this inaptitude, or a means of making the impression strong enough to be retained. The oftener a mental process is repeated, the easier it becomes, and therefore the use of a device for making an impression strong enough to insure its retention by the mind will gradually remove the inaptitude of the mind to receive such an impression.

If we may be pardoned for comparing the mind of a person possessing a weak memory to a block of wood, the operation of this system may be likened to driving a nail into this block. Formulating and writing a question corresponds to the start- ing of the nail ; each successive effort to answer the question corresponds to the effect produced by the blow of a hammer on the nail, which by repeated blows may be driven to its head.

Painstaking effort in the application of this system cannot fail to greatly improve a weak memory. Page after page of a book may be read without producing a lasting impression ; but a question cannot be formulated and written down without leaving an impression of the point it covers, and by repetition and separation the first impressions are necessarily deepened. That which has been learned and remem- bered is filed away in the mind to be produced when required ; while that which is still imperfectly known is studied and re-studied until it also is learned, remem-bered, and filed away.

*Concentration* of the mind and *analysis* of the subject are prerequisites in any system of memory training. The only other element of such a system is a device for insuring *retention*. All such devices are primarily based upon *repetition*. The system herein described, since it covers all these essentials, and insures *concentration*, *analysis*, and *retention*, cannot fail to train the memory and the mind of the student who applies it conscientiously. It is the simplest and most natural system for train-ing the memory, and is therefore believed to surpass in the excellence of its results, the more confusing and complex systems for accomplishing the same result.

# V I.

## LEARNING A BOOK IN ONE READING.

———

The statement that a book may be learned in one reading might properly be questioned if no explanation were made of its exact meaning.

The manner in which a book is to be learned depends upon the object of the reader. If this is to prepare for an ex- haustive examination, great attention must be paid to every point of importance. But the object in general reading is to obtain a general knowledge and the reader is at liberty to choose the points to be specially remembered. Moreover in general reading, many of the important statements in the book are often already known to the reader, having been met with and learned elsewhere. When this is the case, there is a corresponding reduction in the amount remaining to be learned.

It is evident that good judgment is required in determining the salient points of a book. Perhaps no two readers would agree as to what is absolutely essential. A distinguished scholar would no doubt reject much that would be regarded as material by a beginner.

To be of maximum value, a system for learning a book in one reading should be adapted to the use of those who are most in need of assistance.

The system hereinbefore described, while suited to beginners becomes more and more valuable the oftener it is used and trains the mind automatically to grasp the material and reject the immaterial.

A novel is perhaps most thoroughly enjoyed when a reader sits in a capacious

arm chair by an open fire or lazily reclines
on a comfortable couch. Such reading, if
too frequently indulged in, is a mental
dissipation. The impressions on the mind
are light and evanescent. No effort need
be made to deepen or retain them, for as a
rule it is of no importance to remember
what is read.

Books worth learning are frequently
read in the same way, and the mental impressions produced are like "foot-prints
in the sand." It would be palpably
absurd to claim that any system could be
devised for learning a book in one such
reading.

But when a profound student takes up
a treatise on some subject in which he is
deeply interested, his mind gathers the
salient points as a powerful magnet draws
particles of steel from a mass of chaff.
Such a student has no difficulty in learning such a book in one reading. He un-
consciously applies the essential and fundamental principles on which must be based

any practical system for accomplishing a similar result. These are the principles of the system hereinbefore outlined. In this case the mind of the student is trained and retention is assured by the strength of the original impression.

In this kind of reading, the *analysis* consists in determining the essential points which the reader considers it necessary or desirable to remember. Upon coming to such a point, if it is not already known to him, he should formulate a question covering it, and write the question and reference on a card in the usual way. This should be repeated whenever such a point is encountered.

The mental impressions produced by this process may or may not be sufficiently strong to enable the reader upon finishing the book to answer all the questions prepared. If not, the device for deepening the impressions is ready for use and easy of application.

With practice the necessity for the use

of this device will gradually cease ; but it will be an, advantage to have the cards always at hand for reviewing the book at any future time. It is evident that such a review can be most quickly and thoroughly made.

If it is objected that this is not an infallible system for learning a book *in one reading*, it may very truthfully be said that it is as nearly infallible as any other system proposed for accomplishing such a result, and that it is unquestionably the simplest of them all.

The preparation of the questions necessitates the making of a mental abstract of the subject, and the device for learning and remembering this abstract is incomparably simpler and quite as efficacious as any of the complex mnemonic systems sometimes recommended for the same purpose.

# V I I.

# EXTENSION OF THE SYSTEM.

The accumulation of information upon a special subject from any number of different sources is easily and completely effected by using this system. For example, suppose a student is specially interested in a particular subject. In his general reading, he will frequently be struck by some statement bearing on this speciality. This may be an essential point also of the book or article which is being read, or it may be merely an example illustrating such a point.

In the former case write the question covering the statement, with the proper references, on two cards, one of which is to be kept with those pertaining to the book itself, while the other is to be filed under the particular subject to which it pertains. In the latter case, it is only necessary to make out one card for file.

It will sometimes be practicable and advantageous to write the answer on the back of the card.

In this way interesting facts relating to any number of special subjects may be accumulated and filed in the course of general reading ; the cards containing the questions and references being placed in envelopes or pigeon-holes marked with the title of the special subject.

www.ingramcontent.com/pod-product-compliance
Lightning Source LLC
Chambersburg PA
CBHW032117080426
42733CB00008B/974